D0929434

THE FARM

FARMERS

Ann Larkin Hansen
ABDO & Daughters

Published by Abdo & Daughters, 4940 Viking Drive, Suite 622, Edina, Minnesota 55435.

Copyright © 1996 by Abdo Consulting Group, Inc., Pentagon Tower, P.O. Box 36036, Minneapolis, Minnesota 55435 USA. International copyrights reserved in all countries. No part of this book may be reproduced in any form without written permission from the publisher.

Printed in the United States.

Cover Photo credits: Peter Arnold, Inc.
Interior Photo credits: Peter Arnold, Inc.

Edited by Bob Italia

Library of Congress Cataloging-in-Publication Data

Hansen, Ann Larkin.
Farmers / Ann Larkin Hansen
p. cm. -- (The Farm)
Includes index.
Summary: Discusses what farmers do and how their work has changed and continues to change over the years.
ISBN 1-56239-620-X
1. Farmers--Juvenile literature-- 2. Agriculture----Juvenile literature. [1. Farmers. 2. Agriculture. 3. Occupations.] I. Title. II. Series: Hansen, Ann Larkin. Farm.
S519.H355 1996 96-10650
630'.2'03--dc20 CIP
AC

About the author

Ann Larkin Hansen has a degree in history from the University of St. Thomas in St. Paul, Minnesota. She currently lives with her husband and three boys on a farm in northern Wisconsin, where they raise beef cattle, chickens, and assorted other animals.

Contents

79743

Food Comes From Farmers

Farmers are the people who raise the food you eat every day. Some foods, like chocolate and coffee, are grown on farms in faraway countries. Fresh sweet corn may come from a farm just a few miles away. But all our food depends on farmers.

Today, there are less than two million farmers in the United States. Less than one out of every 100 people is a farmer. Each farmer has to raise enough food for more than 100 people.

Opposite page:
A farmer harvesting onions.

Who is a Farmer?

A farmer is anyone who sells more than $1,000 worth of **products** from their farm in one year. A farmer may raise a few **acres** of vegetables, or have thousands of acres of **range** and hundreds of cows. More than one-third of all farmers have other jobs, too!

The average farmer is about 53 years old, and has lived on the farm for more than 10 years. Most are men, but about 150,000 farmers in the United States are women.

A corn farmer near Ames, Iowa.

How Farmers Have Changed

Fifty years ago, about half the people in America lived on farms. Most farmers raised many different things. The farm family had chickens, a pig or two, a big garden, an **orchard**, and some cows and horses. They didn't need to buy much food.

Today's farmers usually raise just one or two crops. Their families buy a lot of groceries at the store. Instead of hiring people to help, farmers use machines to do the work.

Opposite page:
Farm workers pruning grapes in Salinas, California.

Buying and Selling

Though there are fewer farmers, they are raising more food than ever before. In most years, American farmers **produce** much more milk, corn, soybeans, and other food than the United States needs. Much of the extra food is sold to other countries.

To produce so much food, farmers have to buy machines, feed for their animals, **seed**, **chemicals**, gas and oil. Together, farmers spend more than $60 billion dollars each year.

Opposite page:
Combining a wheatfield in Texas.

What Farmers are Like

Farmers come in all shapes and sizes. They are tall, short, fat, thin, nice, or crabby. But all farmers have to be tough. All day long they lift, push and pull. They are on their feet or driving machinery many hours each day.

Farmers are patient. They have to wait for **soil** to dry in the spring, and for rain to come in the summer.

Farmers are good neighbors. They help each other when machinery breaks, or when someone gets sick.

Opposite page: An almond grower in California.

What Farmers Know

Farmers know all sorts of things. They can fix everything from a broken water pipe to a bad starter motor to snapped fence wire. They know when the **soil** is warm enough to plant, and what kind of **seed** to use. They can tell when the weather is right for making hay or picking corn.

Farmers know how to keep baby chicks warm. They know how to give medicine to pigs. They can help a sick cow eat again.

Opposite page: Farmers mulching a field with plastic.

Paperwork and Shoveling Snow

Farmers spend a lot of time studying. They have to read about new kinds of **seeds** and machines. They figure out how much seed they will plant in the spring, how much it will cost, and how much money they will have to borrow.

Farmers have to fill out many forms for the government. They have to figure out taxes. It can take an entire winter to do all this figuring! Many farmers use computers to help with all the paperwork.

Farmers also spend much of the winter shoveling snow.

Opposite page:
A farmer shoveling snow.

Farmers Have Worries

Farmers never know if they will make or lose money each year. A ten-minute hailstorm in a wheat field can destroy months of hard work. A new **disease** can kill half a **herd** of **dairy** cows, and put a farmer deep in **debt**.

Each year farmers worry about floods, **drought**, heat waves, and early frost. They worry about crop prices. If beef or corn prices suddenly fall, many farmers lose all their **profits**. But usually, most farmers make enough money to keep the farm running.

Opposite page:
A farmer in South Carolina.

How Farmers Have Fun

Farmers like to take vacations and have fun just like anyone else. Some farmers fly to Florida in the winter and lay on the beach. Other farmers ask a neighbor to watch their animals and go fishing for a few days.

Some farmers like to hunt deer in the fall after **harvest** is done. Farmers like to go to fairs and **auctions**. Many have hobbies like building **miniature** farm equipment. Some go to the city to shop and visit friends.

Opposite page: A farmer and his grandson at a county fair.

New Kinds of Farmers

As people discover new kinds of food, farmers learn to raise that food. For example, many farmers have learned to raise catfish, trout, and shrimp in large ponds. There are thousands of **organic** farmers who have learned to grow **herbs**, vegetables, grains and meats without **chemicals**.

One of the fastest-growing types of farms is the **greenhouse** and **nursery** business. These farmers raise trees, bushes, flowers, and other plants to sell to city gardeners.

Opposite page:
An organic potato farm.

Caring for the Land

Since there will never be any more land than there is now, farmers must be careful not to wear it out. The dirt must be kept from washing or blowing away. The correct kinds of **fertilizer** and **compost** must be used to keep the **soil** healthy.

Farmers plant trees and bushes to keep the wind from blowing the dirt. They **plow** in certain ways and directions to stop rain from washing soil down hills. Many farmers are using fewer **chemicals** to kill weeds and bugs.

Farmers care for their land.

Farmers Like Farming

Farming is hard, dirty work. There is always something going wrong. The days are long. There is too much paperwork. Farmers don't make much money. So why do farmers keep farming?

Farmers keep farming because they like being outside. They like being their own boss. They like working with plants, animals and machines. They like learning and experimenting. They like the feel of wind in their face and sun on their back. There's nothing on this Earth like farming!

Opposite page: A farmer plowing with a draft horse.

Glossary

acre (AY-ker)—a square piece of land that is 202 feet (61 m) long on each side. Farm sizes are calculated in acres.

auction (AWK-shun)—a sale at a set time and place where items are sold to the highest bidder.

chemical (KEM-ih-kull)—manufactured liquid or powder used to fertilize, or to kill weeds or insects.

compost (KAHM-post)—a rich, crumbly substance made by controlling the decay of plant and animal wastes.

dairy (DARE-ee)—production of milk and related products.

debt (DETT)—owing money to another.

disease (duh-ZES)—a sickness.

drought (DROWT)—a long period without rain.

fertilizer (FUR-tuh-lie-zer)—any substance put on the soil to make it richer and better able to grow crops.

greenhouse—a shed built of glass or clear plastic, kept warm, and used to grow plants.

harvest—bringing in crops.

herb (ERB)—a plant whose leaves and stems are used for medicines and seasoning. Sage and mint are herbs.

herd—a group of one kind of animal, moving together.

miniature (MINN-uh-cher)—small models.

mulch—covering or protecting the ground in order to reduce evaporation and maintain soil temperature.

nursery—gardens used to raise young trees, shrubs, perennials, and seed plants for later use elsewhere.

orchard—a piece of ground on which fruit trees are grown.

organic—using no manufactured chemicals in the production of food.

plow—turning the soil to make ready for planting. Also, the name of the implement used in this process.

produce—fruits and vegetables; or the act of growing or raising food.

product—anything grown or raised on the farm.

profit—the gain from a business; what is left when the cost of goods and carrying on the business is subtracted from the amount of money taken in.

range—open grassland, used for animal pasture.

seed—the grains or ripened ovules of plants used for sowing.

soil—part of the Earth's surface that plants are grown in; dirt.

Index